Good, Occasionally Rhyming

Rob Stepney writes on medicine, science, travel and *The Archers*. He has edited poetry anthologies, co-written two books of poetry, and contributed to BBC Radio 4's *From Our Own Correspondent*.

Kathy Clugston chairs the long-running radio programme *Gardeners' Question Time* on BBC Radio 4. Prior to that she was a continuity announcer for many years and read *The Shipping Forecast* hundreds of times. In 2014 she co-wrote *But First This*, a comic stage musical about Radio 4.

Compiled by
Rob Stepney &
Kathy Clugston

GOOD,

OCCASIONALLY

RHYMING

Å

DK | Penguin
Random
House

First published in the United Kingdom in 2025 by

August Books, an imprint of
Canelo Digital Publishing Limited,
20 Vauxhall Bridge Road,
London SW1V 2SA
United Kingdom

A Penguin Random House Company
The authorised representative in the EEA is Dorling Kindersley Verlag GmbH. Arnulfstr.
124, 80636 Munich, Germany

A CIP catalogue record for this book is available from the British Library.

Print ISBN 978 1 83598 330 0
Ebook ISBN 978 1 83598 343 0

Illustrations by Redgate Arts

Cover design by Sarah Whittaker

Cover images © Shutterstock

Printed and bound in Great Britain by Clays Ltd, Elcograf S.p.A.

Look for more great books at
www.augustbooks.co | www.dk.com

1

To Vice-Admiral Robert FitzRoy FRS, dogged captain of Darwin's HMS Beagle, and pioneer of the systematic forecasting of weather at sea: born 1805; died, by his own hand, 1865. A man who knew how deep depressions form but could not escape the fate inherent in his own.

And also to the many others — whether mariners or forecasters — who work to save lives among those in peril at sea.

Contents

Acknowledgements

This anthology developed from an idea first aired at the 2022 Charlbury Arts Festival, organised by Tony Lloyd. In addition to Rob Stepney and Kathy Clugston, Jim Lee, Ian Richards, Anthony Landale, John Lanyon, Ed Fenton, Adrian Lancini, Kate Smith, Paddy Gallagher, Christina Surawy, Kath Lucas, Geoff Griffiths and the Charlbury Morris took part.

A few years ago, one of *The Spectator*'s poetry competitions had *The Shipping Forecast* as its theme. The large number of entries reflected the status of the Forecast as a national institution. Contributions were full of poignancy and wit and came in varying styles: there were sonnets, even a villan-elle. Poets who have very generously allowed us to include their work are: Alan Millard, Katie Mallett, Bill Greenwell, DA Prince, Max Ross, Brian Allgar, 'Bosun Higgs', Paul Carpenter, Alanna Blake, Chris O'Carroll, Mike Morrison, Joe Houlihan, Sylvia Fairley, Bill Webster, Frank McDonald and Basil Ransome-Davies. We are very grateful both to the individual authors and to *The Spectator* for organising the competition.

Faber and Faber Ltd kindly granted permission to use the following: *Glanmore Sonnets VII* from *Field Work* by Seamus Heaney; *Finisterre* from Sylvia Plath's *Collected Works*; *A Private Bottling* (extract) by Don Paterson, from *God's Gift to Women*; and *Closedown* by Wendy Cope from *Family Values*. Audio rights to the latter were kindly granted by United Agents. Elisabeth Mahoney's *The Poetry of North Utsire* is copyright

Guardian News & Media Ltd 2025 and reproduced by kind permission. Jo Ellison's *The Pure Poetry of the Shipping Forecast* (*Financial Times*, 5th January 2024) is used under licence from the *Financial Times*. All Rights Reserved.

Prayer from *Mean Time* by Carol Ann Duffy (published by Anvil Press Poetry, 1993, copyright © Carol Ann Duffy) is reproduced by permission of the author c/o Rogers, Coleridge & White Ltd, 20 Powis Mews, London W11 1JN. *The Shipping Forecast* extract from *Saturday Night Fry* by Stephen Fry (© Stephen Fry, 2009, published by BBC Radio 4) is reproduced by kind permission of David Higham Associates. *At Smithfield, waiting to get in* by Imtiaz Dharker, from *Over the Moon* (Bloodaxe Books, 2014) is reproduced with kind permission. www.bloodaxebooks.com @bloodaxebooks (twitter/facebook) #bloodaxebooks. David Whyte's *Finisterre*, from his collection *Pilgrim* (© 2012 David Whyte), is reprinted with permission from David Whyte and Many Rivers Company, LLC, Langley, WA. www.davidwhyte.com.

The parody of *The Shipping Forecast* written by the late Les Barker is included by kind permission of his executor, Alma Belles. The parodies from the sketch show *One*, written by David Quantick and Dan Meier, are also published by kind permission. We are also very grateful to Ed Fenton, Adrian Lancini and Desmond O'Connor for generously allowing us to use their work.

Farewell Finisterre by Gillian Clarke (from *A Recipe for Water*, Carcanet Press Ltd, 2009) is reproduced by kind permission of author and publisher, as is Alice Oswald's *Beaufort Poem Scale*: (Copyright © Alice Oswald) reproduced by permission of A M Heath & Co. Ltd. Authors' Agents.

Sean Street's *Shipping Forecast, Donegal* is included by kind permission of the author; and John O'Donnell's *Shipping Forecast (for my father)* by kind permission of the author and The

Dedalus Press, Dublin. Murray Lachlan Young kindly granted us permission to use his poems *The Shipping Forecast* and *The People's Shipping Forecast*.

DA Prince's *Background Music* was first published in *London Grip*, December 2017. *Leaving the World Service* was published in her collection *Common Ground*, Happenstance Press, 2014. Rob Stepney's *Fishing Forecast* was first published in *A Funny Way with Words*, Wychwood Press, 2013. Ed Fenton's *Shipping* and Adrian Lancini's *Now That's What I Call the Shipping Forecast!* were first published in *Words Go Out to Play*, Charlbury Press, 2018. We are also very grateful to Michael Howe (*The Shipping Forecast 00.48*) and Mic Wright (*Occasionally Good: the Forecast from the Unmet Office*) for allowing inclusion of their work.

Each new Shipping Forecast-related event unleashes a fresh wave of creativity. On New Year's Day 2025, it was Radio 4's own celebration of the anniversary of the BBC's first broadcast: several pieces were written especially for it and fresh voices gave new life to familiar phrases. *Shipping Forecast* by Zaffar Kunial was one such poem and we are delighted to be able to include this previously unpublished work. *Shipping Forecast* Copyright © Zaffar Kunial is reproduced by permission of the author c/o Rogers, Coleridge & White Ltd., 20 Powis Mews, London W11 1JN.

Sea Areas and Selected Coastal Weather Stations

Introduction

Rob Stepney

Exact yet evocative, arcane but everyday, fundamentally the same though always different – *The Shipping Forecast* holds a treasured place in our hearts.

Described variously as a litany, mantra or lullaby – and as soothing, enigmatic, measured, and hypnotic – the daily BBC Radio 4 broadcasts provide details about wind, precipitation, pressure and visibility in the waters bordering the British Isles: radio waves that bind us to those at sea. *The Shipping Forecast* is delivered using a precise vocabulary and with all the expressive qualities of a metronome. And the nation loves it.

For all the wildness of the weather often described, 'the Ships' is disciplined. Even with conjunctions, there are barely a hundred possible words. Faced with the Forecast's constraints on language, writers are inspired to challenge its limits; and its solemnity has been eulogised and parodied in equal measure.

This tribute to *The Shipping Forecast* is a collection of poetry and prose diverse in style and content. There is everything from sonnets to a lullaby, with allusions to John Masefield, Dylan Thomas, John Betjeman and Shakespeare. There is punning and pathos, and much of course on weather and the sea as metaphors for the ebb and flow, sunshine and squalls, of human relationships.

A unifying factor is the delight in wordplay, riffing off the obscure and straightjacketed yet somehow inspirational vocabulary of the Forecast itself.

The thirty-one sea areas – drawn like a safety net cast around the British Isles – take you on a virtual voyage starting halfway between the Shetlands and Norway, zig-zagging down the North Sea, scooting westwards along the English Channel, down the coast of France, Spain and Portugal, and then via both sides of Ireland up to the west coast of Scotland and the nearest shard of Iceland.

The list of sea areas itself almost scans and sometimes rhymes: *Humber, Thames, Dover, Wight; Forties, Fisher, German Bight*. Michael Palin says *The Shipping Forecast* is full of poetic possibility. Few would disagree.

No one who has heard it will forget Alan Bennett's melancholy refrain of 'rough or very rough' while reading the forecast as a New Year's treat on the *Today* programme. Equally memorable is the reading in support of *Comic Relief* by John Prescott, old salt and long-time MP for Hull East, in which he insisted his home sea area was *'Umber*, because 'it's the way we say it up there'. And the wide response to *The Spectator*'s competition for Shipping Forecast-related poetry showed that this gem of broadcasting truly belongs not just to mariners but to people as a whole. In a sense, *The Shipping Forecast* is the nation's love poem to the sea, in all its uncertain moods and frequent turbulence.

In the broadcasts, 'warning' is a keynote; 'imminent' a sharp reminder of impending doom: if this is where you are, then this is what you'll get. Batten down the hatches and ride it out. How wonderful to be safe in the sitting room or, even better, snug in bed, with this book.

Reading The Shipping Forecast

Kathy Clugston

Reading the late-night *Shipping Forecast* is one of the most thrilling and, when you first start, terrifying duties of the Radio 4 announcer. It's more than merely reciting a list of places and weather-based words, although doing so clearly and within the allotted time is your primary purpose. It's a ritual, a soother, maybe even a life-saver, and not only for those at sea.

Listeners have often written in – by letter back then, now by email – to say how much they love listening to it. For those overseas it's a reminder of home; for others, a connection to a beloved family member or friend; for many it is, as the poet Imtiaz Dharker put it, 'a calm and unhurried antidote to mess and chaos'. It reminds us that we are not alone. I treasure a letter I received from a listener struggling with anxiety who found that the litany of familiar place names and the soothing tones provided much-needed solace.

What most surprised me when I first got my hands on a Shipping Forecast – or, more accurately, a Shipping Bulletin, which comprises the Shipping Forecast, Weather Reports from Coastal Stations and the Forecast for UK Inshore Waters – was how ordinary it looked. I'm not sure what I was expecting; a document inked on parchment? Flown in by seagull? In fact, a blizzard of tiny black words arrives by email. Everything is in capital letters, with scant punctuation. The sea areas often come in groups followed by the wind direction, wind speed (using the Beaufort scale), precipitation and visibility.

FISHER GERMAN BIGHT
SOUTHWEST 4 TO 6. RAIN. GOOD, OCCASION-
ALLY POOR

The full shipping bulletin takes around eleven minutes to read. A pencil is your best friend: a few dashes here to isolate tricky subdivisions like NORTH NORTH UTSIRE and SOUTH SOUTH EAST ICELAND, an underlining there, a numbering of pages in case you drop them, a noting of the time you want to hit each section so that you reach the national anthem bang on time.

The hundreds of bulletins I read during my time as an announcer were largely incident-free, bar the occasional coughing fit and a couple of occasions when I had to read it twice, either because the previous day's bulletin had been sent by mistake and I hadn't noticed, or due to some technical hitch. My former colleague Peter Jefferson (in his book *And Now The Shipping Forecast*) recalls a rite of passage in the 1970s when colleagues set fire to his script as he was reading it. He somehow managed to get to the end before the flames consumed the paper. How high a pitch his voice reached is not recorded. Even without today's smoke detectors and stricter disciplinary measures, such japes are unlikely as the late announcer now sits in the studio all alone.

Preparation done, you read the credits of the last programme of the day and start *Sailing By* – the music that precedes the late Shipping Bulletin – timing it to finish at exactly 00:48 a.m. The studio lights are turned down low. Your foot taps to a silent beat as you get into the flow. The rhythm is everything. Start too slow and you'll end up gabbling through the last page in a breathless rush; too fast and you'll leave yourself acres of time to fill. You must remember to pause. And breathe.

An actor will try to bring a script to life, to imbue it with emotion – this you must not do. 'Good' at the end of a sentence is not a cause for joy; it is merely a statement about visibility. Neither must you betray sadness at an area of low pressure 'losing its identity', nor excitement at the delicious prospect of 'light icing'. The goal is to be measured without veering into monotony. Tranquil, but not tranquilising. And to leave room for the little bit of magic that has inspired the creative individuals whose work appears in the pages that follow.

And that completes my introduction. Have a safe and peaceful read.

A Short History of The Shipping Forecast

1831: Captain Robert FitzRoy sets sail on HMS *Beagle*. Among those on board is the naturalist Charles Darwin, whose presence was arranged by Francis Beaufort, inventor of the Beaufort scale.

1854: FitzRoy founds the Meteorological Office.

1859: Steam clipper the *Royal Charter* is wrecked off the coast of Anglesey in October's great storm.

1861: FitzRoy issues the first gale warning. Information is sent to the coast by telegraph and a visual system of cones and drums used to warn vessels. The first Public Weather Forecast is published in *The Times*.

1865: FitzRoy dies by suicide. For years, he had suffered episodes of depression. He was also experiencing financial difficulties, having spent much of his fortune on funding meteorological projects.

1911: The first weather forecasts are sent by telegraph to ships at sea.

1914: Forecasts are discontinued at the start of the First World War, resuming in 1921.

1924: The Air Ministry begins a service called *Weather Shipping* transmitted by radio from London. Of the original forecast areas, seven (Forties, Dogger, Thames, Wight, Shannon, Hebrides and Humber) are still in use.

1925: In October, the BBC's Daventry transmitter begins broadcasting the first dedicated marine forecast – at 10:30 a.m. daily, with an evening broadcast added the following year.

1939: Radio forecasts cease during the Second World War.

1954: *Weather Shipping* becomes known as *The Shipping Forecast*.

1956: Heligoland becomes German Bight; the northern part of Forties becomes Viking; Dogger is split in half to create a new area, Fisher; and Iceland is renamed South-East Iceland.

1978: Having been broadcast on the BBC's *National Programme* (until 1939), the *Home Service* and the *Light Programme* (which became Radio 2), *The Shipping Forecast* moves to BBC Radio 4.

1983: The Minches sea area is merged with Hebrides.

1984: North and South Utsire, previously parts of Viking, become separate areas.

1993: On 18th December, the forecast was simultaneously read on Radio 4 (by Laurie Macmillan) and televised on BBC2.

2002: Finisterre is renamed to avoid confusion with a sea area used by the French and Spanish meteorological offices. The name FitzRoy was chosen in honour of the Forecast's creator.

2011: Former Deputy Prime Minister John Prescott (a former merchant seaman) reads *The Shipping Forecast* on Radio 4 for Red Nose Day.

2012: An extract from *The Shipping Forecast* is read during the opening ceremony of the London Olympics.

2014: On 30th May, the 05:20 a.m. forecast was read in Broadcasting House but – for the first time in ninety years – it was not transmitted at the scheduled time. It was eventually transmitted one hour and twenty minutes late.

2024: Radio 4 ceases its Long Wave-only transmissions.

2025: The 100th anniversary of the first Shipping Forecast on BBC Radio.

Stations of the night

Since 1925, *The Shipping Forecast* has been broadcast on BBC national radio. It is a service accessible not only to mariners but to anyone on land who cares to listen, and millions of us do. Many poems reflect this intimate connection between radio waves and weather at sea. It is perhaps most intensely felt by those who tune in late at night or in the early hours.

Farewell Finisterre

Gillian Clarke

One a.m., and I'm alone
with the late night announcer.
We navigate the small hours,
over Viking, the Utsires, his voice
telling the island's rosary,
the stations of the night.

In the house the bottles are empty,
candles snuffed in their lighthouses,
a pulse of flame before dark fell
on the waters of Dogger, German Bight.
His words home along the airwaves.
Humber, he says, Thames, Dover, Wight.

Windows are doused one by one.
The house sleeps beyond Trafalgar, Finisterre.
The wind picks up and my heart's listening
for Lundy, Irish Sea, till words turn north,
his voice saying Shannon, Rockall,
the far away poetry of Hebrides,

and we wing out over the sea where once
from a plane travelling a latitude beyond
Fair Isle, Faroes, South-East Iceland,

I saw far below the coldest word
in the school Atlas. Its arctic radio name.
Its plates of ice. Its silence.

The Pure Poetry of the Shipping Forecast

How a nightly dispatch for sailors became a beloved fixture for landlubbers

Jo Ellison

From the *Financial Times*

It's always seemed one of the more yawning ironies that a nightly radio dispatch designed to protect sailors from the most treacherous stretches of water should have been co-opted by half of its audience as being the equivalent of aural Xanax. Conceived in 1861 by Vice-Admiral Robert FitzRoy, the maritime weather warning was developed following the loss of the *Royal Charter* off Anglesey in 1859 with the loss of 450 lives. Produced by the Met Office, a version of the weather bulletin began in 1924, although it was only broadcast by the BBC in 1925.

The forecast has saved thousands of lives, but its practical application has long been superseded by more precise meteorological and satellite data. Hence, the vast majority of its remaining 6.5 million listeners today are landlubbers, tucked up safe on dry land.

Over time it has become a beloved cultural icon, a tacit expression of national identity, and a nightly liturgy. Poised somewhere between the highest finger beam of moonlight and before the birds start squawking, the forecast is a fixed point for insomniacs, easing anxious souls into a more somn-ambulant world.

As creative writing, the forecast is pure poetry: a project of hard science now enshrined as art. Its exotic roll call has been eulogised by Seamus Heaney — '*Dogger, Rockall, Malin, Irish Sea:/Green, swift upsurges, north Atlantic flux/Conjured by that strong gale-warning voice*'.

It's the place names wherein the real emotions stir. Like something out of Dickens, we imagine fragments of each place: the barren Rockall, Scottish Cromarty, swagged in tartan, enticing Biscay, German Bight. The very strange-ness and old-timey romance of each station conjures a world before Google maps. As is noted in Sanna Nyqvist's essay *Poetics of the Shipping Forecast*, the science of cartography is also a means by which to navigate and reflect upon ourselves. Like examining the Milky Way, listening to the forecast reassures us of a certain permanence in a strange and changing world.

My affinity with the forecast developed as a one-time member of the maritime community, inasmuch as I once 'sailed' the waters of Chelsea while living on a houseboat on the Thames. Our mooring was hitched to one of the city's most prestigious postcodes: I stepped off the boat and on to Cheyne Walk. Our boat was permanently moored there on account of its being about to sink; it leaned heavily to starboard, and I had to rig up a plastic chute around the bed to catch the unnervingly treacle-coloured leaks. In winter it was less well insulated than a cardboard box and the electric wiring around the hull was worryingly exposed.

Even so, for some nine months in my early twenties I considered myself the very height of chic: I could nip into Partridges, the poshest Chelsea grocer, to buy my dinner, and my rent was only £90 a month. During that time I also developed a keen awareness for the vicissitudes of water: far from lulling me to sleep at night, the Thames would slap the wall next to my head. High tide at nighttime became a point of terror as the boat would rock with unprecedented violence. I would marvel that even in the centre of London, one was still completely at the mercy of a greater and more powerful natural force.

At night, listening to the Thames hiss and gurgle, I would wait for the calm of *Sailing By*, the strange little maritime waltz composed by Ronald Binge that would preface the night's Forecast. And while I could never understand a word of the forecast, and all the weather conditions seemed 'moderately poor', I found great comfort in thinking of other people lying in their boat bunks being lurched around with me.

Turns out, I wasn't very shipshape. For years after, I had a recurring nightmare that I was drowning and have never willingly slept on a boat again. Tyne and Dogger, on the other hand, have become my fondest friends. And that sweet, melodious Shipping Forecast will always be my favourite lullaby.

Leaving the World Service

DA Prince

It's still too dark to see the weather
or what the day is wearing. Minutes
are slowly hardening in their shells, setting
into hours ahead. The radio

talks us down, out of that far unknown
of dreams and other countries,
and the light comes on in the fridge
next to the resentful kettle.

It's like a border crossing, where
the night shift waits the dawn relief of headlights,
owls in the valley silent. Or that pause
before the tide gulps at its new direction.

Out on the street a walker, hooded, faceless,
joins up deep shadows, swimming underwater
between the lamp posts; at the last pool of light
he vanishes until tomorrow.

One other car, sharing the early road;
an unnamed bus positioning its start
for the first sweepings of the working day.
Sludge of grit and salt, and that every-morning voice

shipping us out now, measuring the gales,
visibility, strengths, the inshore waters,
anchors for the unknown day
here: rough ground once more called *Home*.

The poetry of North Utsire

Elisabeth Mahoney

From *The Guardian*

As Thursday slipped into Friday, I ended the day as I so often have – on the approach to 1 a.m., put a book down, lights off, prepare to snooze and listen to the FM Shipping Forecast on Radio 4. Over on long wave, the same forecast had just collided with the final wicket of England's Ashes victory over Australia in Sydney. Oops.

Commentators had done their best to warn of the clash. 'It's one of those remote shipping forecasts we can do nothing about,' said one. 'It can't be moved.' As the forecast and simultaneous final wicket grew near, the tone grew more desperate, more disaster. 'There is a shipping forecast heaving into view,' listeners heard, as if a giant liner was approaching a dinghy. 'Try to retune if you're listening on long wave.'

The cricket dinghy had to move out of the way, given that it was available on several other platforms. But there's more to it than that. *The Shipping Forecast* is emphatically not a fixture that hangs about indefinitely for some cricket match to end, however well England might be doing. Those potentially at peril on the sea rely on these bulletins, and Radio 4's change-hating landlubber audiences do too.

Sandwiched between *Sailing By* and the national anthem as the station closes down, the forecasts are one of the network's self-defining gems, and one of its best-loved slots for urban listeners like me, who've only ever been to sea on a holiday pedalo. These forecasts, with their place names, terms (veering, backing) and weatherly detail you never hear in the rest of life, and their hypnotically formulaic progression (area, wind direction, strength, precipitation, sea conditions, visibility), have a talismanic, haunting power.

This is especially the case if you're listening on land, warm and safe, with a hot-water bottle tucked under your feet. From here, the segments of each forecast are like haiku poems: intense, compressed, full of something living and changing, but so still in their composure: 'Viking, North Utsire, westerly, backing southerly, or south-westerly five to seven, perhaps gale eight later, wintry showers, good occasionally poor.'

To listen is a sensual treat, embellished by the velvety voices that read the forecasts, and at the same time an unchanging, formal sign-off to the day. In this context, they can seem like prayers, a litany, especially in the dark if you're half-asleep. If you're wide awake, they have their gently comic bits too: Scilly Automatic still makes me laugh.

I have friends who have named pets after the sea areas, others whose email addresses echo them, and one who called her Edinburgh art gallery Doggerfisher. Writers, musicians, poets and Brian Perkins have all revelled in the forecast's unlikely allure.

But it's unsurprisingly poets who have best captured the feeling of listening to the forecast, perhaps because they too use language that can be very practical in a way that's also magically figurative, beyond the drudge of chatter. Seamus Heaney's sonnet *The Shipping Forecast* focuses on the union of soft voice and strong weather, while Carol Ann Duffy in *Prayer* cherishes the far-off place names heard from home.

Darkness outside. Inside, the radio's prayer/Rockall. Malin. Dogger. Finisterre. The cricket, she might have added, can wait.

Prayer

Carol Ann Duffy

Some days, although we cannot pray, a prayer
utters itself. So, a woman will lift
her head from the sieve of her hands and stare
at the minims sung by a tree, a sudden gift.

Some nights, although we are faithless, the truth
enters our hearts, that small familiar pain;
then a man will stand stock-still, hearing his youth
in the distant Latin chanting of a train.

Pray for us now. Grade 1 piano scales
console the lodger looking out across
a Midlands town. Then dusk, and someone calls
a child's name as though they named their loss.

Darkness outside. Inside, the radio's prayer –
Rockall. Malin. Dogger. Finisterre.

A Private Bottling (excerpt)

Don Paterson

Back in the same room that an hour ago
we had led, lamp by lamp into the darkness
I sit down and turn the radio on low
as the last girl on the planet still awake
reads a dedication to the ships
and puts on a recording of the ocean.

I carefully arrange a chain of nips
in a big fairy ring; in each square glass
the tincture of a failed geography,
its dwindled burns and woodlands, whin-fires, heather,
the sklent of its wind and its salty rain,
the love-worn habits of its working-folk,
the waveform of their speech, and by extension
how they sing, make love, or take a joke.

Rockall-by-Bailey

Alan Millard

Rockall-by-Bailey, on the high seas,
Slip into dreamland and drift on the breeze
From Portland to Plymouth and Lundy to Sole
Where mountainous ocean waves thunder and roll;
Snug in your cot watch the dolphins leap high
On their journey past Shannon to Malin and Skye
Where quarrelsome seagulls and guillemots nest
Under Stornoway's storms sweeping in from the west;
Sleep soundly through Fair Isle, Forties and Tyne,
Deaf to the winds as they whistle and whine,
Blind to the thunder clouds, greyer than slate,
For the tumult will pass and the gales will abate;
Slumber through Humber, Thames, Dover and Wight
Till Portland Bill's lighthouse comes back into sight
And winks as it welcomes you home from the deep.
Rockall-by-Bailey, sleep baby, sleep!

Rocked by the radio waves

Katie Mallett

Humber, Dogger, Thames and Wight
Send me off to sleep each night.
Dover, Portland, Plymouth, Sole
Rock me as the waves that roll.

Fastnet, Lundy, Irish Sea
Provide a lullaby for me,
Whilst Rockall, Malin, Hebrides
Help to put my mind at ease.

Sometimes the Faeroes and Fair Isle
Conspire to give a sleepy smile,
Whilst Bailey, Shannon, Fitzroy shout
They're still around, although far out.

As my mind is shutting down
And deep in Biscay all thoughts drown
I'm drawn far south, Trafalgar way
To dream of sun another day.

Background Music

DA Prince

The backing track to leaving home
whatever the sky says or the weather,
playing its salt notes along Midlands roads,
telling the time in its own measure

better than any clock. Gales in Cromarty, Forth
drum at the traffic lights: you up the volume,
catching Dogger, Fisher as the lights turn green.
Foot down, across the M1, hearing Humber

veering northerly. On time: the road ahead
as empty as a winter sea. Swing past
the hospital as Dover, Wight prepare
for thundery showers. Blue lights at A&E

and at the roundabout Biscay's becoming rough –
the roundabout, landmark, and where you, as ever,
remind yourself Sole isn't where you thought,
till here's a parking spot, the journey over,

the General Synopsis, 05:00 hours,
losing its identity. The day's map drawn
for them, for us; landlocked, a key code and ID
opens the gate, as forecast, as routine.

Recital

Bill Greenwell

Nicely-spoken palpitations
In the early hours of night:
Steadily, like incantations:
Fisher, Dogger, German Bight.
As the sleepless settle in
To the darkness they patrol,
As stealthy as a Bedouin:
Lundy, Fastnet, Shannon, Sole.
A roll, a schoolboy brotherhood,
Uttered to the teacher's liking —
Hoping for the comment, 'Good':
Rockall, Malin, Bailey, Viking.
Perhaps a tribute to the lost,
Now their bitter lives are over —
Quietly, their graves embossed:
Fitzroy, Biscay, Portland, Dover.

The Shipping Forecast (00.48)

Michael Howe

It is a lullaby.
Almost.
You'll be swept to sleep,
Measuring waves and dreaming gales
in every line.

In that absorbing space –
a fretted estuary
between floating and
finally falling,
a pregnant time
to sense the night's stillness and storms
as one wave, one open mouth
to speak the safety of ships
far from land –
you'll sail islands, find channels
through sandbanks, and chart the seas
between what we hear and
what we understand.

Closedown

Wendy Cope

for Alice Arnold[1]

An almost empty building:
Someone, all alone,
Reads the shipping forecast
To a microphone.

Listeners in bedrooms,
Listeners at sea,
Thousands of them, hear her
Speak invisibly.

Hear her through the darkness,
Hear her say goodnight,
Picture her alone there,
Switching off the light.

Is it really like that?
I asked if could go
And be with the announcer

---

[1] Alice Arnold is a former Radio 4 announcer with whom Wendy Cope
spent an evening of research.

In the studio.

And, yes, it's really like that.
Someone all alone,
Reads the shipping forecast
To a microphone,

Speaks into the darkness,
Says a last goodnight,
Plays the national anthem,
Switches off the light.

Of storm and stillness

On 26th October 1859, the *Royal Charter*, a steam clipper bound for Liverpool and almost in sight of home, was one of two hundred vessels lost as the worst storm of the nineteenth century hit our western coasts. The ship, with around five hundred people on board, many carrying Australian gold, was smashed against rocks on the eastern coast of the Isle of Anglesey by hundred-mile-an-hour winds.

The scale of the tragedy, and belief in the possibility of preventing such events, convinced Vice-Admiral Robert FitzRoy that forecasting was needed to protect those in peril at sea. Indeed, he coined the term 'weather forecast'. The Meteorological Office under his leadership systematically collected and disseminated daily weather reports. In February 1861, FitzRoy went a step further and initiated the National Storm Warning Service. The wreck of the *Royal Charter*, and FitzRoy's far-sighted response to it, is the origin of *The Shipping Forecast* as we know it today.

Within two months of the *Royal Charter* disaster, Charles Dickens was in Wales to report on the event and wrote a harrowing description which contrasts the comforts of home with the perils of the sea.

The Shipwreck (excerpt)

Charles Dickens

Abridged from *The Uncommercial Traveller*, 1860

Never had I seen a year going out under quieter circum-stances. Eighteen hundred and fifty-nine had but another day to live, and truly its end was Peace on that sea-shore that morning.

So settled and orderly was everything seaward, in the bright light of the sun and under the transparent shadows of the clouds, that it was hard to imagine the bay otherwise. The tide was on the flow, and had been for some two hours and a half. There was a slight obstruction in the sea within a few yards of my feet: as if the stump of a tree had slipped a little from the land.

O reader, haply turning this page by the fireside at Home, and hearing the night wind rumble in the chimney, that slight obstruction was the uppermost fragment of the Wreck of the Royal Charter, Australian trader and passenger ship, Homeward bound, that struck here on the terrible morning of the twenty-sixth of October, broke into three parts, went down with her treasure of at least five hundred human lives, and has never stirred since! Cast up among the stones and boulders of the beach, were great spars of the lost vessel, and masses of iron twisted by the fury of the sea into the strangest forms. The timber was already bleached and iron rusted, and even these objects did no violence to the prevailing air the whole scene wore, of having been exactly the same for years and years.

Yet, only two short months had gone, since a man, living on the nearest hill-top overlooking the sea, being blown out of bed at about daybreak by the wind that had begun to strip his roof off, and getting upon a ladder with his nearest neighbour to construct some temporary device for keeping his house over his head, saw from the ladder's elevation as he looked down by chance towards the shore, some dark troubled object close in with the land.

He and the other, descending to the beach, and finding the sea mercilessly beating over a great broken ship, had clambered up the stony ways, like staircases without stairs, on which the wild village hangs in little clusters, as fruit hangs on boughs, and had given the alarm. And so, over the hill-slopes, and past the waterfall, and down the gullies where the land drains off into the ocean, the scattered quarrymen and fishermen inhabiting that part of Wales had come running to the dismal sight.

And as they stood, stricken with pity, leaning hard against the wind, their breath and vision often failing as the sleet and spray rushed at them from the ever forming and dissolving mountains of sea, they saw the ship's life-boat put off from one of the heaps of wreck; and first, there were three men in her, and in a moment she capsized, and there were but two; and again, she was struck by a vast mass of water, and there was but one; and again, she was thrown bottom upward, and that one, with his arm struck through the broken planks and waving as if for the help that could never reach him, went down into the deep.

It was the clergyman himself from whom I heard this, while I stood on the shore, looking in his kind wholesome face as it turned to the spot where the boat had been. So tremendous had the force of the sea been when it broke the ship, that it had beaten one great ingot of gold, deep into a strong and heavy piece of her solid iron-work: in which, also, several loose sovereigns that the ingot had swept in before it, had been found, as firmly embedded as though the iron had been liquid when they were forced there. It had been remarked of such bodies come ashore that they had been stunned to death, and not suffocated. Observation, both of the internal change that had been wrought in them, and of their external expression, showed death to have been thus merciful and easy.

The report was brought, while I was holding such discourse on the beach, that no more bodies had come ashore since last night. It began to be very doubtful whether many more would be thrown up, until the north-east winds of the early spring set in. Moreover, a great number of the passengers, and particularly the second-class women-passengers, were in the middle of the ship when she parted, and thus the collapsing wreck would have fallen upon them after yawning open, and would keep them down. A diver made known, even then, that he had come upon the body of a man, and had sought to release it from a great superincumbent weight; but that, finding he could not do so without mutilating the remains, he had left it where it was.

It was the kind and wholesome face then beside me that I had purposed to see when I left home for Wales. I had heard of that clergyman, as having buried many scores of the shipwrecked people; of his having opened his house and heart to their agonised friends; of his having used a most sweet and patient diligence for weeks and weeks, in the performance of the forlornest offices that Man can render to his kind; of his having most tenderly and thoroughly devoted himself to the dead, and to those who were sorrowing for the dead. I had said to myself, 'In the Christmas season of the year, I should like to see that man!' And he had swung the gate of his little garden in coming out to meet me, not half an hour ago.

Finisterre

Sylvia Plath

This was the land's end: the last fingers, knuckled and
    rheumatic,
Cramped on nothing. Black
Admonitory cliffs, and the sea exploding
With no bottom, or anything on the other side of it,
Whitened by the faces of the drowned.
Now it is only gloomy, a dump of rocks —
Leftover soldiers from old, messy wars.
The sea cannons into their ear, but they don't budge.
Other rocks hide their grudges under the water.

The cliffs are edged with trefoils, stars and bells
Such as fingers might embroider, close to death,
Almost too small for the mists to bother with.
The mists are part of the ancient paraphernalia —
Souls, rolled in the doom-noise of the sea.
They bruise the rocks out of existence, then resurrect
    them.
They go up without hope, like sighs.
I walk among them, and they stuff my mouth with cotton.
When they free me, I am beaded with tears.

Our Lady of the Shipwrecked is striding toward the
    horizon,

Her marble skirts blown back in two pink wings.
A marble sailor kneels at her foot distractedly, and at his
     foot
A peasant woman in black
Is praying to the monument of the sailor praying.
Our Lady of the Shipwrecked is three times life size,
Her lips sweet with divinity.
She does not hear what the sailor or the peasant is saying —
She is in love with the beautiful formlessness of the sea.

Gull-coloured laces flap in the sea drafts
Beside the postcard stalls.
The peasants anchor them with conches. One is told:
'These are the pretty trinkets the sea hides,
Little shells made up into necklaces and toy ladies.
They do not come from the Bay of the Dead down there,
But from another place, tropical and blue,
We have never been to.
These are our crêpes. Eat them before they blow cold.'

Rockall

Michael Roberts

Comforting is sleep, but the comfort fails:
The waves break on the bare rock; the traveller remembers
Shipwreck, the struggle in the waters, the wild climb,
Cries of the wind, and then nothing.

Rockall, two hundred miles west of Benbecula,
Bare rock, eighty-three feet wide, seventy feet high,
First seen by Captain Hall, 1810, reported inaccessible –
The last spur on the Great Atlantic Shelf.

How shall the mind think beyond the last abandoned
    islands?
The gulls cry, as they cry in the isles of despair,
The waves break, as they break on Tiree or Foula;
Man is alone; and death is certain.

Was it better to have died in shipwreck?
Here, naked under the bare sky,
The traveller walks, and sanity is the same as madness,
Under the grey sky pressed down to the sea's rim.

Glanmore Sonnets VII (The Shipping Forecast)

Seamus Heaney

Dogger, Rockall, Malin, Irish Sea:
Green, swift upsurges, North Atlantic flux
Conjured by that strong gale-warning voice,
Collapse into a sibilant penumbra.
Midnight and closedown. Sirens of the tundra,
Of eel-road, seal-road, keel-road, whale-road, raise
Their wind-compounded keen behind the baize
And drive the trawlers to the lee of Wicklow.
L'Etoile, Le Guillemot, La Belle Hélène
Nursed their bright names this morning in the bay
That toiled like mortar. It was marvellous
And actual, I said out loud, 'A haven,'
The word deepening, clearing, like the sky
Elsewhere on Minches, Cromarty, The Faroes.

Small Boat Man

Peter Allmond

He never prayed to any god as far as I could tell,
just listened for Portland and Plymouth,
then held his own breath beneath its cliffs
and drift of gulls for the sudden take, the pollock line's
    tight, deep run
tied from his father's blood to the tobacco rope stretched
    against the thumb
for the same blind knife cut through to a touch of skin,
dark spit with the catching tide, dark boat in the early
    hours
dipping him down to a lost sea blue shine of lobster,
knife nicking its fighting claw at the wrist and lifting it up
    for me to see,
the way a father might show his son his whole life,
the old sea, the slow, rhythmic haul from the deep unseen
    before he was.
And now he's gone the village tightens to rock,
a small, coffined boat lowered by rope, settling alongside
    Billing,
Pascoe, Liddicote, men who knew the sea and never
    learned to swim.

Beaufort Poem Scale

Alice Oswald

As I speak (force 1) smoke rises vertically,
Plumed seeds fall in less than ten seconds
And gossamer, perhaps shaken from the soul's hairbrush,
Is seen in the air.

Oh yes (force 2) it's lovely here,
One or two spiders take off
And there are willow seeds in clouds

But I keep feeling (force 3) a scintillation,
As if a southerly light breeze
Was blowing the tips of my thoughts
(force 4) and making my tongue taste strongly of italics

And when I pause it feels different
As if something had entered (force 5) whose hand is lifting
    my page
(force 6) So I want to tell you how a whole tree sways to
    the left
But even as I say so (force 7) a persistent howl is blowing
    my hair horizontal
And even as I speak (force 8) this speaking becomes diffi-
    cult

And now my voice (force 9) like an umbrella shaken inside
    out
No longer shelters me from the fact (force 10)
There is suddenly a winged thing in the house,
Is it the wind?

German Bight

Frank McDonald

Do not go gentle to the German Bight
Rage, rage and rowing keep the sea at bay;
Be like a Viking ready for a fight.

And when you leave the sanctuary of Wight,
The waves will thunder, menacingly grey.
Do not go gentle to the German Bight.

When Fair Isle tempts and even may excite,
Beware the sirens singing far away;
Be like a Viking ready for a fight.

Go forth to meet the demons of the night
And brave gigantic storms where monsters play;
Do not go gentle to the German Bight.

There dragons lurk; a thousand perils invite
And mariners unwary always pay.
Do not go gentle to the German Bight.
Be like a Viking ready for a fight.

Seeing through the night

WJ Webster

The ring of odd and yet familiar names
Recited in its stately, settled round
Beguiles us as a soothing day's-end sound
Whose litany of states and numbers tames
Wild elements with words, and neatly frames
In measured lines those forces which, unbound,
Can render vessels wrecked and sailors drowned
As victims that the challenged ocean claims.
For those at night who brave the open sea
(Not those prepared for sleep in some quiet place)
The forecast, as an overseeing eye,
Keeps watch beyond their own vicinity:
They're tuned to catch the hazards that they face —
Not hear some quaint euphonious lullaby.

Dracula (excerpt)

Bram Stoker

Abridged from *Dracula*, 1897

One of the greatest and suddenest storms on record has just been experienced here, with results both strange and unique. The day was fine till the afternoon, when gossips called attention to a sudden show of 'mares' tails' high in the sky to the north-west. The wind was then blowing from the south-west in the mild degree which in barometrical language is ranked 'No. 2: light breeze'.

The approach of sunset was grand. Before the sun dipped below the mass of Kettleness, its downward way was marked by myriad clouds of every sunset-colour – flame, purple, pink, green, violet, and all the tints of gold; with here and there masses of seemingly absolute blackness.

The wind fell away entirely during the evening, and at midnight there was a sultry heat. There were but few lights at sea. The only sail noticeable was a foreign schooner with all sails set, seemingly going westwards.

Shortly before ten o'clock the stillness grew oppressive, and the silence was so marked that the bleating of a sheep inland or the barking of a dog was distinctly heard. A little after midnight came a strange sound from over the sea, and high overhead the air began to carry a strange, faint, hollow booming.

Then without warning the tempest broke. The waves rose in growing fury, each overtopping its fellow, till in a very few minutes the lately glassy sea was like a roaring and devouring monster. White-crested waves rushed up the shelving cliffs. To add to the dangers, masses of sea-fog came drifting inland – so dank and damp and cold that it needed little effort of imagination to think that the spirits of those lost at sea were touching their living brethren with the clammy hands of death.

On the East Cliff the new searchlight was ready for experiment. The officers in charge got it into working order. Before long the searchlight discovered a schooner with all sails set, the same vessel which had been noticed earlier. The rays of the searchlight were kept fixed on the harbour mouth. The wind suddenly shifted to the north-east, and the remnant of the sea-fog melted; and then, *mirabile dictu*, between the piers, leaping from wave to wave as it rushed at headlong speed, swept the strange schooner before the blast. The searchlight followed her, and a shudder ran through all, for lashed to the helm was a corpse, with drooping head, which swung horribly to and fro at each motion of the ship.

There was a considerable concussion as the vessel drove up on the sand. Every spar, rope, and stay was strained, and some of the 'top-hammer' came crashing down. But, strangest of all, the very instant the shore was touched, an immense dog sprang up on deck from below and jumped from the bow on the sand. Making straight for the steep cliff, it disappeared in the darkness.

Skellig

Rob Stepney

As the second gin and tonic from Heathrow
Smooths the plane's path to the New World,
Look down to see the Old one slip quietly away
In a shrinking line of land, ending in two full stops
Before the unpunctuated Atlantic:
Europe's last lonely rocks

The next time I saw the Skelligs was in a holy roller, post-
    storm swell
That had been born near Boston,
No G&Ts to smooth the small boat's course.
Where Brân the Blessed might have flung a Snowdon
    range into the sea
Rising above the waves, only craggy summit rocks

Great Skellig's seven hundred feet of sandstone grit
Is scaled by steps that creep towards the narrow ledge
Where beehive huts perched above the drop
Cling defiantly like limpets to the rocks

Otherwise, in all of seven centuries,
The monks left only cramped and rudimentary burial
    grounds
Where wind and rain have scoured the headstones clean

Of any clue to why they so loved this savage place
That stinks of gannets
And where only sea pink, campion and scurvy grass
Grow among the rocks

Tortured by the sight of mainland shores
Whipped into submission by lashing rain and wind,
Was it to punish or to purify themselves,
Or just to know the loneliness of God,
That they chose to live and die upon these rocks?

Relation-ships

The Shipping Forecast is, in a sense, a love poem to the sea. It can also convey anxiety and turbulence. In both senses, its description of the weather – experienced and expected – has been used as a metaphor for human relationships, including love and loss

Shipping Forecast

Zaffar Kunial

Had I a shell I'd have put its echoes in your hand.
With words I tried to push your hospital bed. Miles
to the coast. *Mum, that loud oxygen machine sounds
a bit like the sea.* With my ear close as possible
I heard what became your last words. *No rhythm.*
Yours was the first I knew. What could be more inland –
your broader beats, that held mine. Shores. In Birm-
    ingham.
Pulsing coasts. Always your radio was on. My bed

at sea now, eyes on the dark, the walls widen
the way a gale blurs outside in. Sounds at the edge
of language. Edging to weather. Eyekin. Ortat deera.
Outat deera. Ortees. Omatee. Imminent.
Soon. Later. I sense waves, storms, that impossible
safety. Time's drifting island, rising steadily.
Face in sea mist. Mum, is that you? Here we go again –

becoming variable. Veering slowly. Losing identity.

Shipping Forecast, Donegal

Seán Street

They have shared still late October,
but salt stones and a broken tree,
the peeled paint on the lifeboat house
chime with places where the glass falls,
prime sources encountering night's bald predictions.

Everywhere winter edges in,
and now the time is ten to six...
Lightness and weight, air's potentials
pressed into words, implication;
here – on all coasts – listening grows passionately tense.

Fair Isle, Faeroes, South-East Iceland,
North Utsire, South Utsire,
Fisher, German Bight, Tyne, Dogger...
This pattern of names on the sea –
Weather's unlistening geography – paves water.
Beyond the music, the singing
of sounds – this minimal chanting,
this ritual pared to the bone
becomes the cold poetry of information.

The litany edges closer –
Lundy, Fastnet and Irish Sea...

Routine enough, all just routine,
Always his eyes guessing beyond
the headland, she perhaps sleeping, no words spoken.

He stretches forward to grasp it,
claims his radio place – and now
the weather reports from coastal stations
and then: Malin Head – such routine
that she barely glances up, but hears
now falling.

Love Song

DA Prince

Valentia, my sweetest love,
Sandettie's playing jazz above
while we let Ardnamurchan point
the Scilly way to light a joint.
We're in our Forties so we know
how German Bight can spoil the show;
to me your Sole Bay spirit's dearer
than both the kingdoms of Utsire.
My love, Valentia, my dear,
your Biscay's now becoming clear;
the Cape Wrath of our youth is past
and we are Fastnet bound at last.
Let trumpets make the Malin ring
and Rockall dance and Dogger swing.
We'll Lundy on without a care
until we reach our Finisterre.

Finisterre

David Whyte

The road in the end taking the path the sun had taken,
into the western sea, and the moon rising behind you
as you stood where ground turned to ocean: no way
to your future now but the way your shadow could take,
walking before you across water, going where shadows go,
no way to make sense of a world that wouldn't let you pass
except to call an end to the way you had come,
to take out each frayed letter you brought
and light their illumined corners, and to read
them as they drifted through the western light;
to empty your bags; to sort this and to leave that;
to promise what you needed to promise all along,
and to abandon the shoes that had brought you here
right at the water's edge, not because you had given up
but because now, you would find a different way to tread,
and because, through it all, part of you could still walk on,
no matter how, over the waves.

Eros and Faeroes

Paul Carpenter

The nightly shipping forecast
Sends the middle class to sleep,
But not for Miles and Irene
Who've another tryst to keep.

The maritime prognosis
Steers the rhythm of their love
Is it veering, is it backing,
Is it moderate or rough?

From foreplay round the Forties
It's Dogger, Fisher, Bight
And when they've got to Fitzroy
It's time to hold on tight…

Shannon, Rockall, Malin
It must come to a head –
A high that deepens rapidly
Becalms their stormy bed.

Fishing Forecasts

Rob Stepney

Fair Rose
Veering, only moderate
Losing my identity
I am the Minches, merging with Hebrides
Filled with Utsire, north and south
Give me Tyne, Fast net,
Good visibility
Lundy, soul and Fair Isle
Or Ice land squalls

Hi Dogger
Forties, and a Fisher
Losing her identity
It's Dover,
German Blight, and Bailiff,
Ply mouth but Malign head
Rock all!
On the horizon
It's Finis tear.

Shipping

Ed Fenton

Some young people, I hear, are into shipping.
Not all that stuff about tonnage and freight –
It's things like 'Lily ships James' or
'Merlin ships Arthur'. Relationships.
You see, some teenage bloggers are like trainspotters –
Completists, obsessive about facts and figures,
Whether it's Tolkien, Game of Thrones or Doctor Who.
But for other online fans, it's all about the human side…
Why does Remus look at Sirius that way?
It must be love. Love in all its forms –
Tender, at berth, and plain sailing
Love can be a destroyer too
What starts out wavy can soon turn rough
You go through hell and high water, to end up on the rocks
Sometimes the world can seem like one big Bermuda Love
    Triangle
You'd sooner walk the plank than walk down the aisle
Still you've got to take the plunge
You can't spend your whole life contemplating you're naval
Claiming you have a heart of oak
Or waiting for a dreamboat who turns out to be a hulk
You'll get cabin fever
As a child you dreamed of running away to sea
So run away with me

You've got to nail your colours to the mast...
Are you just cruising, frigate! Or is it maritime?

Hebrides, Shebrides

Mike Morrison

Miss Humber Thames-Dover, Miss Humber Thames-
   Dover,
No matter I've traversed the seven seas over –
Wherever I wander, adventure or roam,
The lure of your forecast still beckons me home.

Fair Rose of the Faroes, I honour, respect
Your poignant prose-poetry, clear and correct;
Windspeed and direction console, reassure me
That Somebody, Somewhere will keep a watch o'er me.

O Rockall! My Rockall! My Hebrides, Wight,
Valiant Viking, sweet Sole, Malin Head, German Bight –
I worship you all but in sorrow declare
That I mourn for the Lady of Cape Finisterre.

Inasmuch as I loved you, Cape Wrath, Fitzroy, Forth,
You Utsire Sisters, Ms South and Ms North,
My seafaring's done, it's adieu to the deep;
I'm bound for the Shetland Isles, soundly to sleep.

Becoming Variable

Sylvia Fairley

The sea is calm tonight,
From Tyne to Dogger, Fisher, German Bight;
Light winds in Dover – moderate to good,
Beyond that shingled shore where once we stood
And watched, where white cliffs stand, the ebb and flow
Of tranquil seas, and knew our love would grow.

From Dover – Portland, Lundy, Irish Sea,
Squally, wintry showers. Winds will be
Northerly backing, veering six to eight,
Severe to gale – yet will the seas dictate
The climate of our love – can we appease
The squalls that lash the Outer Hebrides?

Bailey, Fair Isle, Faroes, backing west
To storm force ten – and yet, within my breast
The strength's decreasing moderate to poor,
We've lost the passion that we knew before:
The turbulence is three, becoming slight,
For certitude and joy have taken flight.

Ronald's Way

Max Ross

Ronald my friend's a waspish Wight.
In arguments he'll bark and Bight
Becoming rude and quite Rattray
If it's not going Ronaldsway.

He's never shy at coming Forth
And openly displays his Wrath.
He's worst when he's with Selsey Billy
Whose arguments are just plain Scilly.

The Scots might say: 'Let be Whitby.
Nae need tae get too Cromarty.'
But Ronald is a serious Sole
A Fisherman who'll mope and Mull.
Oddly, his temper nothing Thames
Save shipping news. He loves those names.

Aiming for Cape Wrath

Alanna Blake

I remember, I remember
That voyage to the North
Which took us past the Firth of Clyde
Long after we set forth.

The Mull of Galloway behind
And then the coast of Ayr,
Kintyre, another mull, in front,
The forecast far from fair.

Low cloud hid Ardnamurchan Point,
Cape Wrath still far ahead,
With wild Atlantic waves and sky –
We favoured land instead.

We came ashore on Mallaig's beach
And went back South by train;
Although we planned to try once more
We never met again.

The Forties

Brian Allgar

I turned the wireless up, surprised, last Monday;
They talked about my many long-lost chums
At school back in the Forties: Malin, Lundy,
Cromarty, Fisher, Forth (from Rugby scrums),
FitzRoy, who claimed his name meant royal blood
(We pointed out it meant he was a bastard),
Fat 'Snowball' Wight, who caused the loos to flood,
Young Bailey, who would secretly get plastered.
I chuckled when they mentioned poor old Dogger,
Caught in flagrante with the school nurse, Shannon;
The head, 'Ben' Dover, flogged the filthy snogger,
Expelled him like a bullet from a cannon.
And Tyne, whose father ended on the scaffold;
We called him 'Tyne the Knot' – such callous kids!
But when they came to Faeroes, I was baffled;
Surely they're buried in the pyramids?

The Shipping Forecast

John O'Donnell

For my father

Tied up at the pier in darkened harbour
the two of us below, in cabin's amber
light; me surly in a sleeping-bag, fifteen,
and you, past midnight, calmly tuning in
to the Shipping Forecast, Long Wave's
crackle, hiss, until you find the voice.
What's next for us: rain or fair? There are
warnings of gales in Rockall and Finisterre.
So near now, just this teak bulkhead
between us, and yet so apart, battened
hatches as another low approaches, the high
over Azores as distant as a man is from a boy.
I think of my own boat one day, the deep.
Beside me the sea snores, turns over in its sleep.

The People's Forecast

The Shipping Forecast has been material for a great many homages and parodies: both are ways of showing that we feel it belongs to us all. If love for *The Shipping Forecast* is part of how we see ourselves, it is also part of how others see us. Mary Ellen Foley, who grew up in Kentucky but moved to Surrey, explained in her blog *Anglo-American Experience* that *The Shipping Forecast* is integral to our sense of being a nation of seafarers. Although individuals might find it odd to think of themselves in that way, perhaps in the aggregate she is right. On the *Big Think* website, Frank Jacobs argued that the Forecast is quite possibly the most British thing ever – 'quirkier than cricket, defiantly old-fashioned and ceremonial, as reassuringly regular as Big Ben… and a punctual reminder of Britain's island status.'

So it would have been no surprise to them that the opening ceremony of the 2012 London Olympics was called *Isles of Wonder*, nor that the performance of Elgar's *Nimrod* by the London Symphony Orchestra On Track was overlaid for thirty seconds by a reading of *The Shipping Forecast*, from North Utsire all the way to Wight and Portland.

Even though Stratford-upon-Avon is as far from the sea as you can get, Shakespeare clearly appreciated its significance in the collective psyche. John of Gaunt's poignant but rousing deathbed speech from *Richard II* reflects the importance of the sea in our sense of nationhood.

Bound in with the triumphant sea

William Shakespeare

This royal throne of kings, this sceptred isle,
This earth of majesty, this seat of Mars,
This other Eden, demi-paradise,
This fortress built by Nature for her self...

This precious stone set in a silver sea
Which serves it in the office of a wall
Or as a moat defensive to a house,
Against the envy of less happier lands,
This blessed plot, this earth, this realm, this England...

England, bound in with the triumphant sea,
Whose rocky shore beats back the envious siege
Of watery Neptune ...

The Shipping Forecast

Murray Lachlan Young

Malin cloaks the coast of gales
Hebrides hears ancient tales
From Faeroes where the Selkies swim
South-East Iceland's elven kin
Bailey calls to Baffin Bay
Rockall cries 'Amerikay'
Shannon touches Dingle Bay
Fastnet, Cork, yet Sole might say
Lundy and the Irish Sea
Join Celtic cousins in the lee
Plymouth hears the Breton tongue
Biscay feels the Moorish drum
Fitzroy calls the rolling sea
Trafalgar senses Barbary
Humber holds the Eastern spine
With Cromarty and Forth and Tyne
Fair Isle knows the Viking roar
Utsire's northern shore
Fisher, Dogger, Forties part
Around the North Sea's swirling heart
Above the jaws of German Bight
Thames and Dover, Portland, Wight
Complete a map of landless space
Surrounding this our island race

This storm-tossed, fog-bound alma mater
Veering South-west... occasionally... later.

The People's Shipping Forecast

Collated by Murray Lachlan Young[2]

Autumn, abundant
Apples falling, frequently
Six or seven
Chutney later

School run, fair
Ironing, waiting
Netball later, five

Graduate job search, poor

Paint roller, one or two
Magnolia, increasing slowly

Knitting, storm force nine
Baby soon

Toad in the hole later, three or four

Retiree colonic

---

[2] Radio 4 listeners were invited to use the style of *The Shipping Forecast* to sum up their mood or activities in ten words or fewer. This piece weaves together their contributions.

Weight rising, nine or ten

Dandruff, falling slowly

Anticipation, daughter's birthday, four
Cake rising very, very slowly

Irish Sea, crossing
Sister's funeral
Lows expected
Moderate, later

Chemo, incoming
Cancer, squally
Nausea, rising,
Settling with pills, one or two

Labrador, naughty
Occasionally moderate
Muddy towels later, five or six

York marathon, over sixty
Yesterday's curry
High winds persisting, nine or ten

Brunette, losing identity
Fair, later

Archers' Rob sinisterre
Helen, visibility rising slowly
Severe storm expected

Buttock rash
Going northerly
Doctor's steroid cream

Nightly, one or two

Toddler tired, two or three
Tears expected
Headache, strong to gale force two

Kingfisher sighting
Belfast, sink flooding
Apology accepted
Bifocals found

Tax return deadline
Procrastination, imminent

Dissertation completed
Becoming ecstatic, five or six

Question persistent, nine or ten
Whatever happened
To Finisterre?

This earth, this realm, this forecast

Chris O'Carroll

They measure gales from Viking south to Humber,
Then clockwise round our islands tell each number;
Alert us when there's rain from German Bight
South-west along the continent to Wight;

Announce the visibility today
From Portland to exotic, far Biscay,
Then from Trafalgar north to Irish Sea,
Predict which winds might trend cyclonically;

Warn us of storm clouds rainy, snowy, haily
In western waters, Shannon north to Bailey;
Broadcast conditions of each ocean mile
To South-East Iceland, northwest from Fair Isle;

Of backing winds and veering keep us wise
Via quick stats that sketch our seething skies –
These treasured updates from the BBC,
Like precious stones set in a silver sea.

Doggerel Bank

'Bosun Higgs'

Far from the breezy air, the billows' foam,
Hark to the forecast from our landlocked home!
'Dover and Malin, Sole and German Bight,
Fisher and Viking, Finisterre and Wight.'
Ours the tame life while they in Gale Force 9
Toil without rest in Rockall, Bailey, Tyne
Moderate to good, our outlook here on land.
Deepening lows, or better times at hand
For those that sail the subdivided seas?
Panama flagged, with crews of Congolese,
Plymouth to Dover, East by North you steer
GPS guides you, though winds back or veer;
Wintry the showers in Sole or Ronaldsway –
Tankers, container ships won't stop nor stray;
Clearing to fair, with balmy winds tonight:
So turns what some deem danger to delight.

Occasionally Good: the Forecast from the Unmet Office

Mic Wright

The Shipping Forecast issued by the Unmet Office at 20:25 on Monday 18th May, 2020

We dream of gale warnings
In Plymouth, Biscay, Finisterre, Sole, Lundy
Moments of peace perhaps
From the storm that refuses to shift
Living room, bedroom, bathroom and brain.

The general synopsis is short and ugly:
A Pinter play put on for a handful of punters.

The area forecasts for the next 24 hours:
Viking North pillaging South Utsire.
Forties creeping up slowly, thirties a disappointment.
Cromarty Forth — but didn't Cromarty try hard?
Well done for taking part.
Tyne, Dogging — none of that right now, social distancing
North by North-east, adjusted Hitchcock
Mainly fair, with leading ladies.
German Bight — not as bad as German bark.
North-east 6 or 7, seven ate nine

Rain later. Rain later. Rain.

Humber Thames lifts his heavy body out of the bath,
Veering to the mat, mood decreasing later
Large whisky, occasionally ice
Do not moderate your drinking, occasionally good.
Showers of disapproval.
Portland — remember Bill? He's dead now.
Chance of rain, no chance of regret.
Biscay opens its arms for a hug then remembers. Gale 8
Occasionally severe gale 9.
No honestly, I'm fine.

South Finisterre. North Finisterre — still there, still there.
Sure, where would they go?
Lundy Fastnet — has put her stockings on
She's got notions that one, no better than she thinks she is
Irish Sea — beckons to me:
Come back to Dublin, back to the wind and the rain.
Rockall Malin is doing fuck all today.
Hebrides Bailey looks over at Fair Isle Faeroes,
Waves at South-east Iceland.

All distanced but together.
Occasionally good. Occasionally good.
Occasionally.

At Smithfield, waiting to get in

Imtiaz Dharker

All these girls are waiting
in this city and every city
for something to begin,

holding their thin bodies in their arms,
hissed at by cars that pass
in the rain. They are contained

behind the barricade that draws
a metal line between them
and the freezing vans.

At the meat market across the road,
busy men in white coats are dancing
their daily load of carcasses

into patient rows.
Later in the night their coats
will be smeared with blood.

Later in the night
when Sailing By is done
and the shipping forecast has begun

thinking of all those souls

out in the dark and cold, thinking
of the ones alone, the others

lying side by side, holding hands,
I remember the young girls
who are younger every day

the ragged line they make,
how their legs are blue
and their faces

lit up before they reach
the light inside,
in anticipation of the dance.

When cast away on *Desert Island Discs*, Imtiaz Dharker chose
to bring a recording of *The Shipping Forecast* read by Luke
Tuddenham and including a bit of Ronald Binge's *Sailing By*.

Cruise Blues

Katie Mallett

I must go down to the sea again, for I'm booked up on a
    cruise
Around the whole of Britain for the culture and the views.
While driving to the port I'll have the radio turned on so
I can hear the shipping forecast to inform me as I go.

I must go down to the sea again, although the forecast tells
Of storms from Thames to Dogger with fearsome winds
    and swells,
And then from there to Viking and Fair Isle in the north
There will be snow with blizzards for those who venture
    forth.

From Hebrides to Malin the winds may drop a while,
But right across the Irish Sea fog covers mile on mile,
Going down to Lundy and Fastnet in the west
The visibility will rise, but still not of the best.

Plymouth will be drizzly, but Portland into Wight
Will clear a little, but by then it will be dead of night.
And the channel across Dover will be choppy so they say –
Oh, how I wish I'd booked this cruise for another sunnier
    day.

Pissed again

Basil Ransome-Davies

The Skipper sank another rum and stared into the night.
'Is this the Hebrides?' he asked, 'Or just the German
Bight?'
The First Mate poured himself a tot and answered, 'Don't
ask me,
For all I know it's South Utsire or the Irish Sea.'

They summoned up the boatswain, who'd been at the
bootleg gin
And suggested 'South-East Iceland' with a disrespectful
grin,
Then fiddled with the radio as though it were a toy.
They tipped the numbskull overboard and called the cabin
boy.

He, young lad, was a simpleton. He stank of rotgut wine.
No flicker of intelligence, of morals not a sign.
He mumbled, 'Dogger Fisher — either that or Dover Sole.'
He went into the briny with a kind of Western roll.

The Captain and his Number One took equal turns to
pour
As wicked winds whipped up the waves and battered
Britain's shore.

Both pissed as newts, they slumbered as the ship went
    round and round.
You don't need navigation when you don't care where
    you're bound.

The Shopping Forecast

Joe Houlihan

The general synopsis at 1100:
Shopping trip with spouse, new low expected.
Iceland, Tesco, Aldi
Fare: good.

Co-op, Budgens, Lidl
Variable, paying later.
Bejam, Somerfield
Losing their identities. Very rough.

Waitrose
Becoming tetchy, backing to eatery.
Harrods
Iconic. Occasional pain and misery.

Selfridges, Liberty, John Lewis
Sale warning. Wintry glowers.
Debenhams, Monsoon
Poor, becoming very poor.

The Shipping Forecast

Les Barker

...And now time for the shipping forecast and reports from coastal stations. Here is the general synopsis at 07:00 GMT. Cow in sea area Shannon, moving slowly eastwards and filling. Sorry, that should be Low in sea area Shannon.

And now the area reports: Viking, North Utsire, South Utsire, East Utsire, West Utsire, South-West Utsire and North-North-East Utsire: wind south-west, rain at times, good. Forties, fifties, sixties, Tyne, Dogger, German Bight, French Kiss and Swiss Roll: westerly, becoming cyclonic, good.

Humber, Thames, Bedford, Leyland-DAF, Dover Sole, Hake, Halibut and Monkfish: regular outbreaks of wind, rain at times, good. Wight, Portland, Plymouth, Ginger Rogers and Finisterre: light flatulence, some rain, very good.

Lundy, Fundy, Sundy and Mundy: wind south-west, becoming cyclonic, bloody marvellous.

Rockall: sod all wind, heavy showers, absolutely incredible.

Malin, Hebrides, Bailey, Fair Isle, Cardigan, Pullover and South-East Iceland: wind south-east, rain at times, slightly disappointing. And now the reports for coastal stations:

Tiree: wind north-west, 7 miles, one thousand and four, rising slowly.

Butt of Lewis: north, 5 miles, one thousand and six, falling.

Wolverhampton: north-west, as far as the ring road, nine nine eight, rising slowly.

Norway: nil points.

In 2003, Radio 4 Announcer Brian Perkins recorded this version of *The Shipping Forecast* by the folk poet Les Barker for *Guide Cats for the Blind*, a series of CDs released to help raise funds for The British Computer Association of the Blind.

A selection of parodies from the Radio 4 sketch show One

David Quantick and Dan Maier

And now with the time coming up to ten years later, the *Brit Pop Forecast*:

Oasis. Don't Look Back in Anger. Easterly.

Supergrass. Young, clean, running green.

Sleeper. What was that one? Fading rapidly.

Blur. This is a low. Appropriate. Ironically.

Manic Street Preachers. Still going, amazingly.

Menswear. Reforming. Gales of laughter later.

–

And now with the time approaching 7 o'clock, it's time for the *Faux Pas Forecast*:

When's the baby due? Northerly, five.

This room will look nice when it's decorated. Westerly 5 or 6.

Are you going to Kay's party? Oh haven't you? Um. Well. Maybe she… Anyway. Um. Face becoming red.

Oh yes. Oh yes. That's it. Oh god, yes Laura. I mean Helen. Mortified, becoming awkward later on.

–

And now with the time approaching 5 p.m., it's time for the *Mid-Life Crisis Forecast*:

Forties. Restless: three or four.

Marriage: stale; becoming suffocating.

Sportscar, jeans and T-shirt; westerly, five.

Waitress; blonde; 19 or 20.

Converse All-Stars; haircut; earring; children;

becoming embarrassed.

Tail between legs; atmosphere frosty;

Spare room: five or six.

From Saturday Night Fry

Stephen Fry

And now, before the news and weather, here is the Shipping Forecast issued by the Meteorological Office at 1400 hours Greenwich Mean Time.

Finisterre, Dogger, Rockall, Bailey: no.

Wednesday, variable, imminent, super.

South Utsire, North Utsire, Sheerness, Foulness, Eliot Ness:

If you will, often, eminent, 447, 22 yards, touchdown, stupidly.

Malin, Hebrides, Shetland, Jersey, Fair Isle, Turtle-Neck, Tank Top, Courtelle:

Blowy, quite misty, sea sickness. Not many fish around, come home, veering suggestively.

That was the Shipping Forecast for 1700 hours, Wednesday 18 August.

The Shipping Forecast in song

Rob Stepney

As a readily recognised element of national life, *The Shipping Forecast* is part of popular culture. Even so, the number of songs (of all genres) referring to the Forecast and its sea areas – either directly, or as sampled material for the background to tracks – is surprisingly large.

One of the best examples is Blur's 'This is a Low', from the 1994 album *Parklife*, which plays on the ambiguity between meteorology and mood. Apparently, it was Alex James' Christmas gift to Damon Albarn of a handkerchief map of the sea areas that suggested the song. And so we have lyrics which rhyme 'Bay of Biscay' with 'back for tea' and 'Dogger bank' with 'taxi rank'. As for the Queen, she's 'gone round the bend' and 'jumped off Land's End'. But the overall message – despite there being a low – is that (as the radio says) it won't hurt you: 'When you are alone/ It will be there with you/ Finding ways to stay solo'.

In 'In Limbo', Radiohead wrote of having 'a message I can't read' in Lundy, Fastnet, Irish Sea. 'I'm lost at sea... I've lost my way', the lyrics continue. In 'Dry the River', New Ceremony speak of the angel of doubt which 'crept into your bed' being countered by dancing to *The Shipping Forecast*. And the Young Punx' 2006 energetic 'Rockall' samples Radio 4 announcer Alan Smith reading a real broadcast while the lyrics urge every compass direction to 'Rockall... ROCKALL!' 'by midnight tonight'.

The song 'Stranded', apparently initially written by the Incredible String Band's Mike Heron, was recorded on the 1980 album *Chance* by Manfred Mann's Earth Band. An extract from *The Shipping Forecast* is used at the start, end, and between verses: there is mist from Ronaldsway to Malin Head and rising pressure in Valentia. Since the stranding mentioned by the title seems to be that of a motorcar in Iowa, maritime conditions were probably not responsible. But an American blogger and former DJ, for whom the track is a favourite, records that 'a hypnotic form of weather data is voiced pleasantly by a British woman', which is a fair description.

Chumbawamba, best known for the line 'I get knocked down but I get up again' from their hit 'Tubthumping', refer to *The Shipping Forecast* at the start of 'The Good Ship Lifestyle', and even include a few notes of 'Sailing By' at the end. Given that the band is described as anarcho-punk, is this association parody, heartwarming, or plain odd? Whatever the reason, we have the memorable lines 'Fairless, Bailey, Pharaoh, Hepperties' and 'Travelga, Fiddister, Irisy, Fiscay'.

'Pharoahs', a mellow instrumental from Tears for Fears, has sampled warnings of gales in almost every sea area bar the Faeroes. Thomas Dolby's 'Windpower' ends with an extract, but so faint that it is almost subliminal. Beck's ten-minute 'The Horrible Fanfare' has a Shipping Forecast sample halfway through, as does Jethro Tull's 'North Sea Oil', from the Scotland-inspired album *Stormwatch* – where sampling 'the Ships' is at least merited by context.

But the relevance of the Forecast is clearest in two songs dealing directly with peril at sea. A snatch of broadcast is used in the lament 'The Fishermen's Song' performed by the Scottish folk group Silly Wizard, in which a woman stands by the storm-torn shoreline 'to condemn that wild ocean/ for the murderous loss of her man'.

Gale warnings from Humber through sea areas south and west and as far north as Iceland introduce Justin Sullivan's 'Ocean Rising'. This song (by a founder member of New Model Army) tells of 'a forty-foot wall of water crashing down' on the sailors of the open lifeboat, the *James Caird*. Happily, all survived the astonishing voyage from Ernest Shackleton's ice-bound *Endurance* to South Georgia. So this is a tale of the Southern Ocean. But hazards at sea are the same the world over.

The folk band The Longest Johns came up with an amusing Shipping Forecast parody, incorporating conventional and internet-related references, as a bonus track on their 2018 album *Between Wind and Water*. After the usual preamble, we find: 'Viking. Arriving soon, leaving later with your wife…'. The sea areas are parodied as Fair Isle, Faeroes, Pyramid, Sphinx… and 'Rockall, Sole. One fish, two fish. Red fish, blue fish.'

In the realm of classical music, Cecilia McDowall composed the wonderful 'Shipping Forecast' (2011) for the Portsmouth Festival Choir. It interweaves the words of Seán Street's poem *Shipping Forecast, Donegal* ('the pattern of names on the sea') with those from Psalm 107 ('He maketh the storm calm, so that the waves thereof are still') and from *The Shipping Forecast* itself.

But it is perhaps the account given by Andy White, Northern Irish poet and singer-songwriter, which shows most movingly how *The Shipping Forecast* sails its way into unexpected, somehow magical musical spaces. Andy was looking for something to complete the track 'The Whole Love Story'. He recalls:

'I'd listen to *The Shipping Forecast* late at night… in Belfast, and then up the Northern Highway in County Antrim. It was part of coming home from London to a place where the weather and the sea mattered. I didn't wish I was a fisherman, but I felt kinship in trying to decode the poetry held in the mysterious list of names. Sometimes there was a man's voice, sometimes a woman's.

'There were recognisable places – South of Iceland, Dogger – but very little was clear. The deep and unknowable twins, North and South Utsire, reigned supreme. There were nearly always showers. It was nearly always moderate, and then it was good. All this was poetic information upon which to act. I didn't know when I would use it, but I knew that one day I would.

'Something was missing in the final mix of *The Whole Love Story*. It was written in that half-world of sensation, emotion, fictional autobiography and true-life romance where love songs often live. We had recorded it live in the studio, but it needed more voices. Or another voice.

'I ventured out with a cassette recorder and asked a group of Randalstown schoolgirls eating buns in the bakery if they would repeat the title line into the microphone. They were laughing and spilling crumbs. It was good, but the song still wasn't finished.

'There was to be a storm that night. Everyone on Main Street was talking about it. *The Shipping Forecast* was coming up. We lined up the mix of *The Whole Love Story* and plugged the radio into the desk, rolled the tape to where the lead vocal invites the guitar "tell it to me" and pressed record. I didn't know what the forecast would be, only that the weather would *be* the story in the song – and vice versa.

'I had been hoping it would be a woman's voice. It was. We ran it live onto the recording: the list of enigmatic names building up, the gale approaching. Then the word that sealed the deal, *cyclonic*. The whole love story.'

Musician-ship has also celebrated the Forecast in two rousing pieces, which follow. Adrian Lancini found echoes of sea areas in the lyrics of a dozen pop songs. (Though he had to be persuaded not to rhyme Utsire with desire.) The second piece, which returns to our starting point of radio, is a true singalong sea shanty: *The Shipping Forecast* by Kathy Clugston and Desmond O'Connor from their stage show *But First This: A Musical Homage to Radio 4.*

Now That's What I Call the Shipping Forecast! Volume 1

Adrian Lancini

I can see Cromarty now the rain has gone
I can see only barnacles in my way

Look for the bare Hebrides
The simple bare Hebrides
Forget about your worries and your strife

Ooooooh, I love to love you Bailey
Ooooooh, I love to love you Bailey

Gimme, gimme, gimme Shannon after midnight
Won't somebody help me chase the seagulls away?
Gimme, gimme, gimme Shannon after midnight
Take me over Fastnet to the break of the day

Pull up to the Humber baby
In your mini submarine
Pull up to the Humber baby
And ride into Grimsby

And now I wanna be in Dogger
And now I wanna be in Dogger

I love Rockall

So put another wine in the coolbox baby
I love Rockall
Take the boat off the line and sail with me

I'm sittin' on a dock of Biscay
Watching the tide roll away
I'm just sittin' on a dock of Biscay
Far from Tyne

I'm a Sole Man
I'm a Sole Man

German Bight, German Bight fever
We know how to row it
German Bight, German Bight fever
With Fisher just above it

It's fun to stray to the VIKING SEA AIR
It's fun to stray to the VIKING SEA AIR
It has everything for young men to enjoy
You can hang out with all the buoys

BUT

(Tell me why) I don't like Lundy
(Tell me why) I don't like Lundy
(Tell me why) I don't like, I don't like
(Tell me why) I don't like Lundy
I wanna shoo-oo-oo-oo-oo-oooot
The whole zone down

Britain's Lullaby

Kathy Clugston and Desmond O'Connor

Another long, hard day is done
And into bed you creep
Hoping to progress
Into sweet unconsciousness
But if snoozin' is elusive
And wide awake you keep –
Put on the Shipping Forecast
You're guaranteed to sleep!

And it's German Bight and Humber
That's the way to gentle slumber
By Forth, Tyne and Dogger
You'll be sleeping like a log
And it's Scilly Automatic
It makes Rockall sense to try
To axe the Shipping Forecast
It's Britain's lullaby

The late announcers' voices
Flow gently, soft and slow
I rarely get past Forties
'Fore my eyelids start to go
You can keep your pills and ointments
Forget your counting sheep

It's Fitzroy, Sole and Lundy
That's the remedy for sleep

And it's German Bight and Humber
That's the way to gentle slumber
By Forth, Tyne and Dogger
You'll be sleeping like a log
And it's Scilly Automatic
It makes Rockall sense to try
To axe the Shipping Forecast
It's Britain's lullaby

And it's Portland, Plymouth, Biscay
There's no point in gettin' frisk-ay
Cos by Irish Sea and Shannon
You'll be snoozin' with abandon
It's Scilly Automatic
It makes Rockall sense to try
To axe the Shipping Forecast
It's Britain's lullaby
It's Britain's lullaby

Reflections on (some) sea areas

Rob Stepney

Dogger

Dogger Bank was once 'Doggerland', a land bridge to continental Europe and home ten thousand years ago to Mesolithic people whose tools are occasionally dredged from the seabed.

Slap in the middle of the North Sea, Dogger now is surely not the area most visited, though it is frequently flown over. If Forties, directly to the north, is thick with oil rigs, Dogger is the go-to place for non-carbon energy generation. The fact that much of the sea here is less than 100 feet deep makes it ideal for wind farms, and one ambitious set of plans would make it the world's largest offshore site. Sea area Dogger is popular in the sense that – along with Finisterre – it is the most frequently mentioned in our collection. Perhaps this is because the name is intriguingly difficult to fathom, despite the shallowness of the water.

International relations have played out, sometimes stormily, around our shores. The Dogger Bank has been the site of several naval battles and one 'incident', in October 1904, which fell short of the criterion for a battle since – although several sailors died – only one side was armed.

The armed side was the Russian Imperial Navy, lately out of Baltic ports, and bound for the Far East, where they hoped to join a fleet fighting the Japanese. On the other side were fisherman from Hull who had nothing but a good catch in their sights.

Greater loss of life was avoided only by poor Russian gunnery. The battleship *Oryol* is said to have fired five hundred shells without hitting anything.

In mitigation, it was a bit foggy around Dogger at the time, and there *was* the very faint possibility that the Japanese navy had slipped three-quarters of the way around the world without being noticed. On the other hand, the Dogger Bank is only a hop and a skip away from Hull, and a routine source of fish and chip suppers. So trawlers were not unexpected.

The Dogger Bank encounters that were real battles included those between the British and Dutch in 1781, and between the British and German navies in 1915.

German Bight

Named after 'bucht' meaning 'bay'. From 1949 to 1955, the sea area was known as Heligoland after the minute German islands thirty miles north of the German coast which – remarkably – belonged to Britain from 1807 to 1890 when Heligoland (population a little over a thousand) was swapped for Zanzibar (population 1.3 million), and was again controlled by Britain from 1945 until 1952.

Portland, Wight: when weather forecasts won the day

From Robert FitzRoy onwards, the purpose of *The Shipping Forecast* has been saving lives. Sometimes this is achieved in unexpected ways. Never was the Forecast more important than in June 1944. The vital question was whether the seas would be stormy or still.

The greatest maritime invasion force ever amassed was gathered on the south coast of England, waiting to cross the Channel to land on the beaches of German-occupied Normandy. Embarking troops were concentrated in Hampshire and Dorset, and many landing ships were to rendezvous at a point south of the Isle of Wight.

Five thousand vessels carrying more than 150,000 troops would be involved in the D-Day landings. For success they needed calm seas and clear skies – ideally at the time of a full moon that would aid the preparatory paratroop drops.

The planned date was 5th June, but meteorologists led by the RAF's James Stagg rightly predicted that the weather would be bad, leading to the postponement of the invasion until the following day, when there was likely to be a brief window of more favourable weather. And so it proved. Although conditions were not ideal on 6th June, they would have been far worse on the next date that had been pencilled in.

A key contributor to Stagg's forecast for Channel weather on June 6th was an observation made a few days earlier and far to the west, on the Irish coast of sea area Rockall. There, Blacksod Point's postmistress, Maureen Sweeney, collected hourly weather data. Unknown to her, they were being forwarded to Allied HQ in England.

During Maureen's night shift on June 3rd, her barometer measured a rapid drop in pressure, suggesting the imminent arrival of a storm. An unusual phone call from England asked her to confirm the reading, which she did. And the invasion was postponed. As forecast, the originally scheduled D-day of June 5th was one of potentially disastrous high winds and stormy seas.

Throughout the Second World War, Allied dominance in the North Atlantic deprived German forecasters of timely information on the slew of weather systems approaching from the west. The Germans expected prolonged stormy conditions in June 1944, and many troops normally defending the

Normandy coast, including Field Marshal Rommel, were engaged elsewhere or allowed on leave.

German forecasters also failed to predict the exceptionally harsh winter that contributed to the defeat of Hitler's invasion of Russia. Such broad-brush forecasting remains exceptionally difficult. But – in the competition to achieve the fine-grain forecasting of daily weather – history may well have been fundamentally changed by the predictions for sea areas Portland and Wight on 5–6th June 1944.

Finisterre becomes FitzRoy

There was a bit of a storm, and an RIP, when Finisterre spawned FitzRoy. In 2002, under high pressure from international agencies concerned that the UK's sea area Finisterre did not coincide exactly with areas of the same name used by Spanish and French weathermen, our Finisterre was renamed FitzRoy.

This was not without good reason, given FitzRoy's role in founding the forecast. But his mid-nineteenth-century dream was way ahead of technology's ability to deliver it. He spent the latter part of his life and all of his fortune in a vain attempt to make the impossible happen. When this failed, and after a lifelong history of depression, he killed himself in 1865.

Even though the renaming of sea area Finisterre as FitzRoy was entirely appropriate, it was lamented by some. And these comments can be found on the web. One man wrote: 'RIP Finisterre. A renowned friend of sailors, born in 1949 of Latin extraction. In finer times, "moderate" or "good". In times of sadness, "poor", with unsettling episodes of "veering". The funeral will be held at sea.'

David Brankley said, 'When I was a boy I used to think that the newsreader was saying "finished there". Now I suppose it is.' While 'Jeremy' said simply 'I'll miss Finisterre – but it's not the end of the world.'

Fastnet

South-westerly severe gale force 9, increasing storm force 10 imminent. This warning was part of *The Shipping Forecast* broadcast at 22:30 GMT on Monday 13th August 1979. Hearing it must have sent shivers down the spines of those taking part in the Fastnet ocean race, which had begun in calm seas two days before.

Held every two years, the race requires yachts to sail from Cowes to Plymouth. As the crow flies, it is a modest 142 miles. The catch is that in the process the yachts must round the Fastnet Rock – making the distance more than 600 miles.

Fastnet is not only far away; its weather can be fierce. Eight miles from Cork, the lighthouse in 2017 recorded the highest wind speed – 119 mph – ever experienced around the Irish coast. Recreational sailing into the path of low-pressure Atlantic systems is clearly not without its thrills, but it can be fatal. The 1979 race led to the deaths of eighteen people. Fifteen were yachtsmen and three rescuers.

That said, it is a disturbing fact that, even in the most dangerous seas, the destructive consequences of deliberate human action match or exceed those wrought by nature. It is an invidious record, but Fastnet has seen the greatest loss of life in any sea area. In May 1915, the German submarine U-20 sank the Cunard liner *Lusitania* off the southern coast of Ireland on its voyage from New York to Liverpool. The liner was carrying ammunition for the British war effort, but also 1,266 civilian passengers.

Including members of the crew, 1,195 people died. This loss is almost double that caused by the infamous *Royal Charter* storm of 1859, which sank ships all around the western coast of Britain.

There are deaths at sea, but there are also other kinds of loss. The Celtic coastlines that feature to such an extent in *The Shipping Forecast* are coastlines of emigration. The port of Cork was the single most important point of embarkation for Irish people bound for the USA – 2.5 million passed through it looking for new lives in America.

Shannon

A millennium and a half ago, Irish monks built a monastery in the Atlantic. Still clinging to the mass of rock that is Great Skellig, it is exactly as they left it – and the most striking feature of sea area Shannon, which is named after the estuary of Ireland's longest river, but also includes Bantry and Galway Bays.

The filming of part of *Star Wars: The Force Awakens* on the Skelligs has massively increased the market for trips, and their expense. Twenty years ago, when we visited, it was all very low-key. The day before, I'd phoned the village to see if a trip would be possible and whether our young children could be taken safely to the island. No problem, I was told. And they'll have life jackets.

So, soon after leaving the shelter of Ballinskelligs Bay, when the little boat had started to experience the giant swell, I asked the captain if we could have the life jackets. 'We have them, but we don't recommend you put them on,' he said. Why not? I wondered innocently. 'If someone had worn one yesterday, they might have been sick in it, and then where would we be?' he replied. We agreed that our two-year-old son, at least, should put his on. But it turned out the boat had only adult jackets. 'It'll be all right,' said the captain. 'Just don't blow it up too far'.

And so we crossed to Great Skellig. Eva, aged four, was singing, 'Miss Polly had a dolly that was sick, sick, sick.' My wife was looking the part. But Nick Strong-stomach, aged six, was dreaming he was a Viking sent to pillage the monasteries of western Ireland – which the Vikings did rather a lot, despite being told that they should piss off back to their own sea area.

Rockall

A sheer granite tooth sticking sixty feet above the Atlantic 190 miles west of Scotland. No more than twenty people have set foot on it, and none has endured its isolation for more than 45 days.

Decoding The Shipping Forecast

Kathy Clugston

Sailing By: where days end and the Forecast begins

The iconic tune that precedes the 00:48 Shipping Forecast was written in 1963 by the English composer Ronald Binge (1910–79) for the BBC to use as library music. It's one of his many recognisable melodies, including the popular *Elizabethan Serenade* composed 12 years earlier in 1951, for which he won an Ivor Novello Award. *Sailing By* – with its evocative flute arpeggios and swelling strings – was chosen to accompany the late Shipping Bulletin in 1973 and is still played every night. The tune was the subject of a 2025 New Year's Day edition of *Soul Music* on Radio 4 and was chosen on *Desert Island Discs* by the poet Imtiaz Dharker and the singers Michael Ball and Jarvis Cocker, the latter describing it as 'an aid to restful sleep'.

The recording used by the BBC is performed by the Alan Perry/William Gardner Orchestra and its full duration is 2 minutes and 36 seconds.

Structure

The Shipping Bulletins are made up of three sections. The first part is *The Shipping Forecast* itself, which begins with the incantation: 'And now the Shipping Forecast, issued by the

Met Office on behalf of the Maritime and Coastguard Agency at...' followed by the time, day and date. This is followed by any gale warnings and a general synopsis describing the location of low- and high-pressure areas along with their barometric pressure and likely evolution, for example:

> HIGH SOLE 1022 DISSIPATING BY MIDNIGHT TONIGHT. NEW LOW EXPECTED NORTH FORTIES 1006 BY SAME TIME.

After that comes the familiar list of sea areas starting with Viking and moving clockwise around the map to end with South-East Iceland. Areas with similar conditions are grouped together and the information given in the following order: wind direction and force, weather conditions and visibility (see Glossary for terms used), for example:

> FITZROY SOLE LUNDY WEST 6 TO GALE 8 VEERING NORTH 4 TO 6. RAIN THEN SHOWERS. MODERATE OR POOR BECOMING GOOD.

The Shipping Forecast takes about 3 minutes to read and is also broadcast on Radio 4 at 17:54 at weekends.

The second part of the bulletin is the Weather Reports from Coastal Stations, which loops from Tiree Automatic in the inner Inner Hebrides back round to Machrihanish Automatic. (Automatic weather stations are unstaffed, often in remote locations, and use weather sensors and data loggers to collect and transmit information.) The format here is slightly different; visibility is stated in nautical miles and followed by the barometric pressure and pressure trend:

TIREE AUTOMATIC WEST FOUR
RECENT DRIZZLE 16 MILES 1004 RISING
SLOWLY

The final section is the 24-hour forecast for the Inshore
Waters of Great Britain and Northern Ireland starting with
the general weather situation and then describing conditions
by section of coast in the following order:

Cape Wrath to Rattray Head including Orkney; Rattray
Head to Berwick-upon-Tweed; Berwick to Whitby; Whitby
to Gibraltar Point; Gibraltar Point to North Foreland; North
Foreland to Selsey Bill; Selsey Bill to Lyme Regis; Lyme Regis
to Land's End including the Isles of Scilly; Land's End to St
David's Head including the Bristol Channel; St David's Head
to Great Orme Head including St George's Channel; Great
Orme Head to Mull of Galloway; Isle of Man; Lough Foyle
to Carlingford Lough; Mull of Galloway to Mull of Kintyre
including the Firth of Clyde and the North Channel; Mull
of Kintyre to Ardnamurchan Point; Ardnamurchan Point to
Cape Wrath; Shetland Isles.

The Met office uses a supercomputing system that can
obtain 215 billion weather observations from all over the
world every day. However, at the time of writing, the
expertise of an actual human meteorologist is still required to
compile shipping bulletins every six hours for dissemination
in various forms.

Glossary

The terms used in *The Shipping Forecast* have highly specific meanings.

Timing

Imminent: Expected within 6 hours
Soon: Expected within 6–12 hours
Later: Expected, but not for at least 12 hours

Visibility

Very poor: Less than 1,000 metres
Poor: 1,000 metres to 2 nautical miles
Moderate: 2–5 nautical miles
Good: More than 5 nautical miles

[Note: a nautical mile is a little under 1.2 land miles]

Wind direction

Variable: will change direction
Becoming cyclonic: a considerable change in wind direction across the path of a depression
Veering: direction is changing clockwise, e.g. south-west to west
Backing: direction is changing anticlockwise, e.g. south-east to east

Wind speed

Wind force is measured using the Beaufort Scale from 0 (Calm) to 12 (Hurricane). A gale warning is issued for winds of Gale Force 8 and above:

Gale 8: winds of 29–46 mph, twigs break off trees
Severe gale 9: 47–54 mph, chimney pots and slates torn off
Storm 10: 55–63 mph, trees uprooted
Violent storm 11: 64–72 mph, widespread damage
Hurricane force 12: more than 73 mph, devastation [Note: not an actual hurricane, which is a tropical cyclone.]

Atmospheric pressure

High or low pressure systems can move:

Slowly: at less than 15 knots
Steadily: at 15 to 25 knots
Rather quickly: at 25 to 35 knots
Rapidly: at 35 to 45 knots
Very rapidly: at more than 45 knots

Sea State

Smooth: a wave height of less than 0.5 metres
Slight: 0.5 to 1.25 m
Moderate: 1.25 to 2.5 m
Rough: 2.5 to 4 m
Very rough: 4–6 m
High: 6–9 m
Very high: 9–14 m
Phenomenal: More than 14 m

Bibliography

Toby Carr and Katie Carr, *Moderate, Becoming Good Later*, Summersdale, 2023

Meg Clothier, *The Shipping Forecast: Celebrating 100 Years*, Ebury, 2024

Peter Collyer, *Rain Later, Good*, Adlard Coles Nautical, 1998

Nic Compton, *The Shipping Forecast: A miscellany*, BBC Books, 2016

Charlie Connelly, *Attention All Shipping*, Abacus, 2005

Peter Jefferson, *And now the Shipping Forecast*, UIT Cambridge Ltd, 2011

Peter Nichols, *Evolution's Captain: The Tragic Fate of Robert FitzRoy*, Profile, 2003

Mark Power, *The Shipping Forecast*, GOST, 2023 (expanded edition)